JAKE AND JENNIE TALK ABOUT DINOSAURS

BY JULIE R. TUCKER

Copyright

In accordance with the U.S. Copyright Act of 1976, the scanning, uploading, and electronic sharing of any part of this book without the permission of the publisher constitute unlawful piracy and theft of the author's intellectual property.

All rights reserved. No part of this publication may be reproduced or transmitted in any form or by any means, electronic or mechanical, including photocopying, recording, or by any information storage and retrieval system, without permission in writing from
the copyright owner.

ISBN-13: 978-1533367693

ISBN-10: 1533367698

© Julie R. Tucker 2016

FOR THE
TWO PEOPLE
I LOVE MOST
IN THIS WORLD

~ALAN AND TREY~

Table of Contents

Chapter 1		Dinosaur Facts
Chapter 2		Carnivores – Meat Eaters
Chapter 3		Herbivores – Plant Eaters
Chapter 4		Omnivores – Meat & Plant Eaters
Chapter 5		Armored, Horned, and Plated Dinosaurs
Chapter 6		Largest Dinosaurs
Chapter 7		What Happened to the Dinosaurs? Extinction

More Books from Julie R. Tucker

Chapter 1

Dinosaur Facts

Jake and Jennie are at school and they are talking to Mr. Johnson, the science teacher, about dinosaurs. Mr. Johnson knows a lot about dinosaurs. He says they are very interesting.

Jennie asked Mr. Johnson, "What is a dinosaur?"

Mr. Johnson replied, "The word dinosaur means, terrible lizard. That is basically what a dinosaur is, a big scary reptile."

"How long ago did dinosaurs live?" Jake asked.

"Dinosaurs roamed Earth from 250 million years ago until about 65 million years ago.

There are three main groups of dinosaurs. Carnivores, herbivores, and omnivores." Mr. Johnson explained.

Jennie beamed, "Can you tell us more about dinosaurs and what those words mean?"

"Sure Jennie, we'll start with the carnivores," replied Mr. Johnson.

Chapter 2

Carnivores, Meat Eaters

Mr. Johnson continued, "Carnivores are meat eaters. They had long, strong legs so they could run after their prey to catch it. They also had sharp teeth and strong jaws to eat their food."

"Oh, that is gross!" cried Jennie.

"Cool! What else did they do to get their food?!" shouted Jake.

"Well, they had very good eyesight and a keen sense of smell. Meat eaters had a large brain so they could plan different ways to catch their food.

Some of them traveled in packs or groups so they could trick their prey. But, some of them were also scavengers. Scavengers are animals that find food they did not kill themselves.

Tyrannosaurs Rex or T-Rex and Velociraptors are two of the most popular meat eaters," explained Mr. Johnson.

"Tell us about a different group of dinosaurs, Mr. Johnson," added Jennie.

"Okay, I'll tell you about the herbivores next," replied Mr. Johnson.

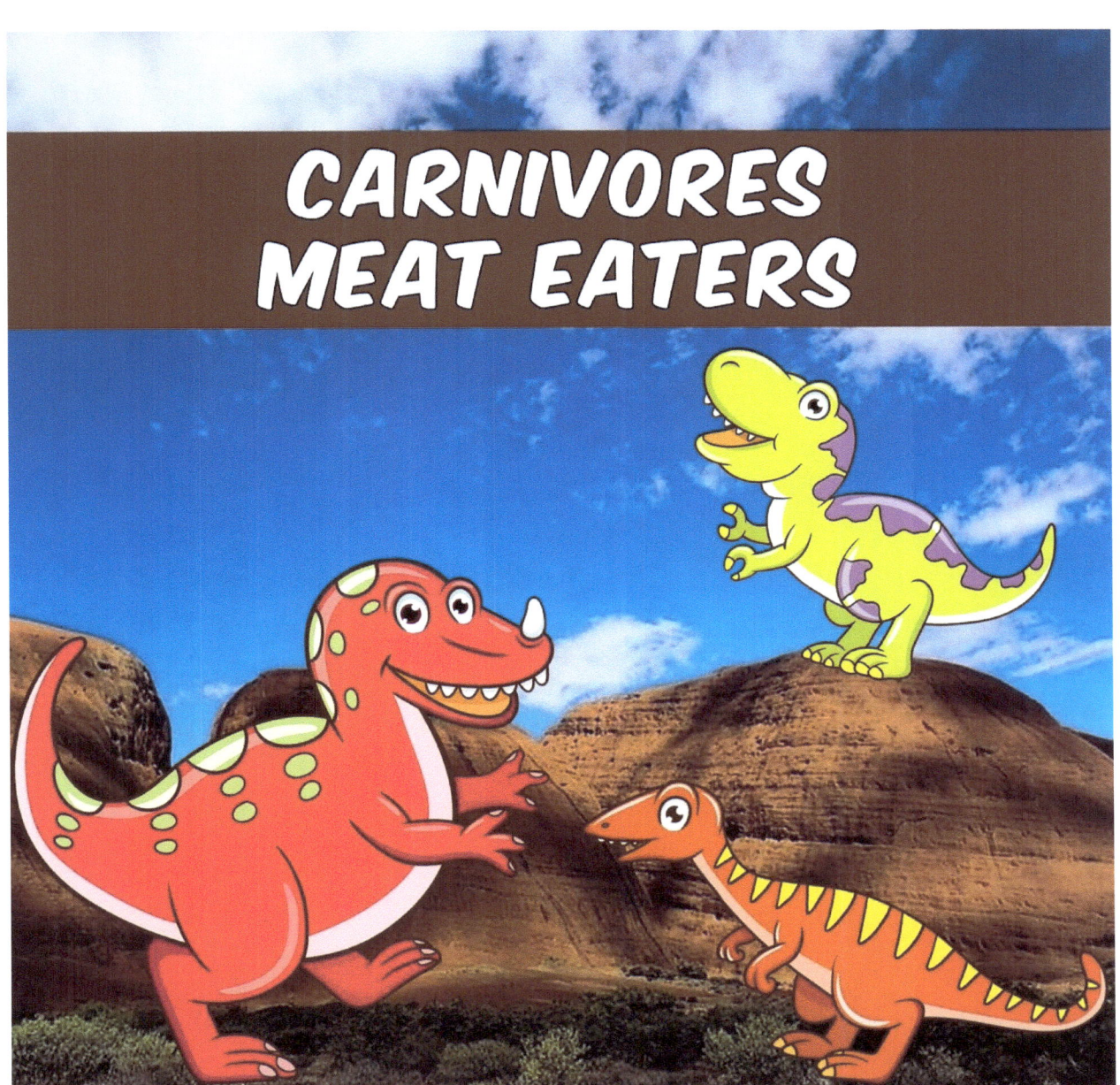

Chapter 3

Herbivores, Plant Eaters

Mr. Johnson began, "Herbivores are plant eaters. Herbivores have blunt, straight teeth that are good for stripping leaves and twigs off of trees. Some of them have flat teeth for grinding hard plants. Many herbivores have pouches in their cheeks to store food for later."

"How much food does a plant eater dinosaur eat every day?" asked Jake.

"Herbivores have to eat a lot more food than meat eaters do. That's because plants don't fill an animal up as much as meat. Plant eaters usually have larger stomachs than meat eaters too. Some plant eaters even swallowed rocks to help grind the food in their stomachs," explained Mr. Johnson.

"What are some names of herbivores?" questioned Jake.

"Alamosaurus, Anchiceratops, and Apatosaurus or Brontosaurus," Mr. Johnson said.

Jennie asked, "Can you tell us about omnivores?"

HERBIVORES PLANT EATERS

Chapter 4
Omnivores, Meat and Plant Eaters

"I will be happy to tell you about omnivores Jennie!" Mr. Johnson happily said.

"There are only a few dinosaurs that were omnivores. They ate both plants and small animals. They also ate eggs and insects. Two omnivore dinosaurs are Ornithomimus and Oviraptor.

It is very easy for omnivores to find food because they can eat almost any food that is around them. They might find fish in the water or even insects on the plants they are eating," nodded Mr. Johnson.

"Are there other types of dinosaurs, Mr. Johnson?" quizzed Jake.

Chapter 5

Armored, Horned, and Plated Dinosaurs

"Jennie and Jake, there were some cool dinosaurs that had armor, horns, and plates! They roamed Earth between 70 million years ago and 137 million years ago. All of these dinosaurs were plant eaters.

One of the armored dinosaurs is called Ankylosaurus. They were one of the last dinosaurs of to live. Their tails were like a heavy hammer with sharp spikes. They used their tails to break the legs of their enemies. Even their eyelids were plated.

The most famous horned dinosaur was the Anchiceratops. They lived about 72 million years ago in North America. They had large necks that were topped with small horns. They had two long eyebrow horns and a small horn on their snouts.

Then there were dinosaurs that had plates on their backs. They are called Stegosaurs. They had double rows of plates and spikes along their backs. They also had sharp spikes on the ends of their tails to protect them from predators.

Now Jake and Jennie let me tell you about the largest dinosaurs that ever walked on the planet!" Mr. Johnson said excitedly.

ARMORED, HORNED, & PLATED DINOSAURS

Chapter 6
The Largest Dinosaurs

Mr. Johnson informed Jake and Jennie, "The largest dinosaurs ever discovered are called sauropods. They are massive and tall, with long necks, and tiny heads! These awesome giants roamed Earth for 140 million years.

They have super long necks for their body size. Dinosaur fossils of about seven species have been found dating back some 160 million years ago.

The biggest dinosaur ever found lived in what we now know as Argentina. The people who found the dinosaur measured the length and circumference of the dinosaur's thigh bone. They estimated it was 65 feet high with a total length of 130 feet. It weighed 77 tons or over 150,000 pounds. That's as much as 14 African Elephants weigh. That's one big animal, don't you think?!"

"That's gigantic!" shouted Jake.

"Yeah, super gigantic!" yelled Jennie

"I have one more thing about dinosaurs I want to tell you. Let me ask both of you, what do you think happened to all of the dinosaurs?" Mr. Johnson asked.

Chapter 7

What Happened to the Dinosaurs? Extinction

"Did they run out of food and get too hungry?" asked Jennie.

"My big brother said a giant meteorite crashed into Earth and that's why they died!" Jake shouted.

Mr. Johnson explained, "Both of those reasons could be correct. We don't know the exact reasons they became extinct, but we have some good ideas.

Jennie you could be right; they may have run out of food because the climate changed. That caused plants to die so they didn't have enough food. When the plant eaters died the meat eaters didn't have any food either.

Jake you could be right; it is believed a giant meteorite crashed into Earth. This caused the climate to change and the dinosaurs could not survive.

Another reason could be, there was ash and gas coming out of volcanoes and that suffocated the dinosaurs. Or diseases could have wiped them out.

The dinosaurs became extinct about 65 million years ago. There are many different beliefs why the dinosaurs died. Any or all of them could be correct.

Just recently, there were four new types of dinosaur fossils found in Japan. There are new discoveries found all the time. All we can do is continue learning about the dinosaurs and hopefully someday we will know all of the answers," explained Mr. Johnson.

"Thank you for telling us about dinosaurs, Mr. Johnson!" beamed Jake.

"Thank you, Mr. Johnson, it was so exciting!" Jennie said happily.

More books from Julie R. Tucker

Thank you so much for reading Jake and Jennie Talk about Dinosaurs. I would love to know if you and your children liked the book and if they learned anything new.

Please let me know by visiting my author's page at Amazon.com/Author/JulieTucker and make a comment or two.

You are welcome to contact me at Julie@JoyfulStorybooks.com

Currently, there are three other books in the series, Fun with Friends, for you to read.

Sam and Susie Talk about Zoo Animals is the first book in the series.
The second book is Max and Maxine Talk about Outer Space.
The third book is Jake and Jennie Talk about Dinosaurs.
Fourth in the series is Kylie, Kate, and Cody Talk about Money.

There is one more book that is not included in the series. This book is for younger reader's ages one through six. It's Zoo Animal Alphabet and Baby Animal Numbers.

Credit

Anchor Me Designs
For the glitter letters in my logo.
Find Marisa Landy at
AnchorMeDesigns.com

www.ingramcontent.com/pod-product-compliance
Lightning Source LLC
Chambersburg PA
CBHW050911180526
45159CB00007B/2879